Summer Solutions.

Minutes a Day-Mastery for a Lifetime!

Intermediate

A

Mathematics

Nancy McGraw

Bright Ideas Press, LLC
Cleveland, Ohio

Summer Solutions Intermediate A
Mathematics

Printed in the United States of America

ISBN 13: 978-1-934210-26-0
ISBN 10: 1-934210-26-9

Cover Design: Dan Mazzola
Editor: Kimberly A. Dambrogio

Instructions for Parents/Guardians

- *Summer Solutions* is an extension of the *Simple Solutions* Approach being used by thousands of children in schools across the United States.

- The 30 Lessons included in each workbook are meant to review and reinforce the skills learned in the grade level just completed.

- The program is designed to be used 3 days per week for 10 weeks to ensure retention.

- Completing the book all at one time defeats the purpose of sustained practice over the summer break.

- Each book contains answers for each lesson.

- Each book also contains a "Who Knows?" drill and *Help Pages* which list vocabulary, solved examples, formulas, and measurement conversions.

- Lessons should be checked immediately for optimal feedback. Items that were difficult for students or done incorrectly should be resolved to ensure mastery.

- Adjust the use of the book to fit vacations. More lessons may have to be completed during the weeks before or following a family vacation.

Summer Solutions Intermediate A

Reviewed Skills

- Place Value Through Hundred Millions
- Addition and Subtraction of Decimals
- Prime and Composite Numbers
- Prime Factorization
- Addition and Subtraction of Fractions and Mixed Numbers
- Multiplication and Division of Fractions
- Multiplication and Division of Decimals
- Read and Write Decimals
- Ratio and Proportion
- Perimeter and Volume
- Area of a Parallelogram
- Area of a Triangle
- Circumference of a Circle
- Math Vocabulary
- Simple Measurement and Conversions
- Problem Solving

Help Pages begin on page 63.

Answers to Lessons begin on page 79.

Lesson #1

1.   A five-sided shape is a(n) _____.

2. $3,615 \times 5 = ?$

3.   Find the mode of 76, 32, 24, 76 and 50.

4.   Which digit is in the tenths place in *45.0873*?

5. $845 \div 25 = ?$

6.   Round *827,693,051* to the nearest hundred million.

7.   How many degrees make up a straight angle?

8. $19,864 + 38,296 = ?$

9.   Rewrite $\dfrac{14}{21}$ in simplest form.

10. $0.7 - 0.2246 = ?$

11.   Find $\dfrac{2}{7}$ of 49.

12. $4.55 \div 0.5 = ?$

13.   List the factors of 18.

14. $0.006 \times 0.04 = ?$

15.   Is *17* a prime or a composite number?

16.   Write *nine and four thousandths* in standard decimal form.

17. $56\dfrac{1}{8} + 38\dfrac{1}{5} = ?$

18.   Figures with the same size and shape are _____.

19.   Put these decimals in order from least to greatest.

        0.6   0.175   0.158   0.044

20.   Jason earned $75 at work last week. He put one-fifth of his earnings in a
        savings account. How much did he save?

1.	2.	3.	4.
5.	6.	7.	8.
9.	10.	11.	12.
13.	14.	15.	16.
17.	18.	19.	20.

Lesson #2

1. $53 \times 22 = ?$

2. Write $9\frac{5}{7}$ as an improper fraction.

3. Which is greater, 6 tons or 8,000 pounds?

4. Draw 2 congruent squares.

5. $6\frac{3}{4} + 3\frac{1}{8} = ?$

6. Which factors of 12 are also factors of 18?

7. $3.21 \times 0.4 = ?$

8. $10\frac{2}{5} - 4\frac{1}{4} = ?$

9. Write the formula for finding the perimeter of a regular polygon.

10. Find the average of 310, 421 and 424.

11. Which digit is in the thousandths place in 2.913?

12. If $3x = 12$, what is the value of x?

13. $\frac{5}{8} \times \frac{16}{25} = ?$

14. Closed figures made up of line segments are _____.

15. $31\frac{1}{5} - 17\frac{4}{5} = ?$

16. The temperature at 7 a.m. was 32°F. At 3 p.m., it was 40° warmer; by 8 p.m. it had fallen by 15°. What was the temperature at 8 p.m.?

17. Is this a slide, a rotation, or a reflection?

18. Round 72.841 to the nearest hundredth.

19. What do you call the distance across a circle through the center?

20. $6.34 \div 0.02 = ?$

1.	2.	3.	4.
5.	6.	7.	8.
9.	10.	11.	12.
13.	14.	15.	16.
17.	18.	19.	20.

Lesson #3

1. Write *13.045* in words.

2. $30,000 - 12,648 = ?$

3. What time was it 7 hours and 15 minutes ago, if it is 6:15 now?

4. How many years are in 3 decades?

5. Round *487,853,210* to the nearest ten million.

6. $62.84 + 8.66 + 23.54 = ?$

7. Put these decimals in order from least to greatest.

    15.32   15.032   15.65   15.065

8. Round *76.813* to the nearest hundredth.

9. $2\dfrac{1}{2} \times 3\dfrac{1}{10} = ?$

10. Find the GCF of 12 and 16.

11. Find $\dfrac{3}{5}$ of 30.

12. $3.53 \times 0.03 = ?$

13. $\dfrac{8}{10} \div \dfrac{2}{10} = ?$

14. $85.65 \div 0.05 = ?$

15. Figures having the same shape, but different size are _____.

16. $76\dfrac{1}{6} - 38\dfrac{5}{6} = ?$

17. **This shape is a parallelogram.** It has 2 sets of parallel sides. Draw a parallelogram in the box.

18. Which digit is in the thousandths place in *2.001*?

19. $\dfrac{5}{8} \bigcirc \dfrac{3}{7}$

20. If a bear weighs 345 pounds 5 ounces, how many ounces does it weigh?

1.	2.	3.	4.
5.	6.	7.	8.
9.	10.	11.	12.
13.	14.	15.	16.
17.	18.	19.	20.

Lesson #4

1. Draw a parallelogram.

2. Make a factor tree for *36*.

3. $28,916 + 37,865 = ?$

4. How many years are in 7 centuries?

5. Draw a right angle. How many degrees are in a right angle?

6. Write *5.2071* using words.

7. $\dfrac{9}{10} \div \dfrac{3}{10} = ?$

8. Round *55.936* to the nearest tenth.

9. Write $\dfrac{17}{2}$ as a mixed number.

10. $2\dfrac{1}{6} \times 3\dfrac{1}{3} = ?$

11. $60,000 - 32,817 = ?$

12. How many inches are in 5 yards?

13. $7.14 \times 0.3 = ?$

14. $1.44 \div 6 = ?$

15. Find the volume of this cube.

    7 mm

16. Find the LCM of 10 and 14.

17. $\dfrac{3}{4} + \dfrac{2}{5} = ?$

18. Half of the diameter is called the _____.

19. $0.477 \bigcirc 0.5$

20. Daniel bought a TV for $495, a DVD-player for $59, and 5 DVD's that cost $12.95 each. What is the <u>estimated</u> total for Daniel's purchases?

1.	2.	3.	4.
5.	6.	7.	8.
9.	10.	11.	12.
13.	14.	15.	16.
17.	18.	19.	20.

Lesson #5

1.   All four-sided shapes are called _____.

2.   Put these decimals in order from greatest to least.

     2.45   2.045   2.363   2.303

3.   In a bag of marbles, 5 are blue, 2 are white, 4 are red, and 1 is black. What is the probability of picking a blue one? Red? Green?

4. $36,918 - 19,575 = ?$

5. $15\frac{1}{9} - 6\frac{8}{9} = ?$

6.   How many quarts are in 3 gallons?

7. $1.816 \bigcirc 2.3$

8. $\frac{5}{6} \times \frac{18}{25} = ?$

9. $2.22 \times 0.04 = ?$

10. 7 tablespoons contain how many teaspoons?

11. Round *26,743,985* to the nearest thousand.

12. Figures with the same size and the same shape are _____.

13. Which digit is in the hundredths place in *16.814*?

14. $0.9 - 0.5561 = ?$

15. Find the area of this parallelogram.

16. $49.07 \div 7 = ?$

17. $765 + 388 = ?$

18. Write the decimal number for *ten and three thousandths*.

19. Karen worked $5\frac{1}{2}$ hours on Monday, $4\frac{1}{2}$ hours on Tuesday, and $6\frac{1}{4}$ hours on Wednesday. How many total hours did Karen work?

20. $93 \times 26 = ?$

1.	2.	3.	4.
5.	6.	7.	8.
9.	10.	11.	12.
13.	14.	15.	16.
17.	18.	19.	20.

Lesson #6

1.   How many cups are in 5 pints?

2. $2\frac{2}{3} + 2\frac{1}{6} = ?$

3.   What temperature is shown on the thermometer?

4. $6.65 + 0.7 + 5.0 = ?$

5. $13 - 4\frac{2}{7} = ?$

6. $3{,}888 \div 36 = ?$

7.   Is *61* a prime or a composite number?

8.   What is the probability of getting a 5 on one roll of a die?

9. $42.36 \div 0.6 = ?$

10. Find the area of the parallelogram.

11. $0.006 \times 0.008 = ?$

12. If $5x = 45$, what is the value of x?

13. $\frac{5}{7} \times \frac{14}{20} = ?$

14. Find the GCF of 10 and 25.

15. $24{,}806 \times 3 = ?$

16. Make a factor tree for *50*.

17. Write $\frac{84}{9}$ as a mixed number.

18. Write the formula for finding the volume of a prism.

19. Draw an obtuse angle. Does it measure more or less than 90°?

20. To park in the city parking lot costs $1.75 for the first hour. It costs $.75 for each half hour after that. How much did Mr. Morgan have to pay to park in the lot for $4\frac{1}{2}$ hours?

1.	2.	3.	4.
5.	6.	7.	8.
9.	10.	11.	12.
13.	14.	15.	16.
17.	18.	19.	20.

Lesson #7

1. What is the name for *the number that occurs most often* in a set of data?

2. Find the LCM of 12 and 16.

3. $\dfrac{5}{8} + \dfrac{3}{8} = ?$

4. Find $\dfrac{3}{5}$ of 20.

5. 0.625 ◯ 0.75

6. $\dfrac{6}{8} \div \dfrac{2}{8} = ?$

7. How many centimeters are in 12 meters?

8. Write the formula for finding the area of a parallelogram.

9. Find the perimeter and the area of this rectangle.

18 in.

5 in.

10. Write $9\dfrac{3}{7}$ as an improper fraction.

11. Name this shape.

12. $\dfrac{2}{3} \times \dfrac{9}{10} = ?$

13. Write *twenty and twelve hundredths* in decimal form.

14. $35.62 \times 0.03 = ?$

15. On the Fahrenheit temperature scale, water boils at _____.

16. On the Celsius temperature scale, water boils at _____.

17. Which digit is in the ten thousandths place in *38.9071*?

For items 18 – 20, identify each as a slide, a rotation or a reflection.

18. ⇨ ⇧   19. ⇨ ⇨   20. ⇨ ⇦

1.	2.	3.	4.
5.	6.	7.	8.
9.	10.	11.	12.
13.	14.	15.	16.
17.	18.	19.	20.

Lesson #8

1. $38.96 \div 0.02 = ?$

2.   The area of a square is 64 mm². What is the length of each side?

3. $43.6 + 9.28 + 0.362 = ?$

4.   Find the average and the mode of 52, 34, 62, 23, 71 and 52.

5.   Write the formula for finding the area of a rectangle.

6.   Give the estimated difference of 926 and 392.

7.   Name this shape.

8.   Find $\frac{2}{5}$ of 55.

9.   0.4 $\bigcirc$ 0.19

10.   Write the decimal number for *thirty-seven and twenty-two hundredths*.

11.   What is the probability of the spinner landing on a number less than 3?

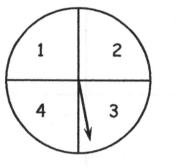

12.   Find the LCM of 4, 6 and 12.

13.   Round *15.081* to the nearest tenth.

14.   How many feet are in 5 miles?

15. $38\frac{3}{5} + 19\frac{1}{2} = ?$

16. $0.05 \times 0.07 = ?$

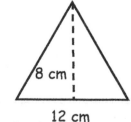

17.   Find the area of the triangle. (See the *Help Pages* for examples.)

18.   Draw a parallelogram.

19.   What do we call the bottom number in a fraction?

20.   In November, Mr. Torrez sold 28 cars during the 1st week, 42 cars during the 2nd week, and 56 cars during the 3rd week. At this rate, how many cars will he sell during the 1st week of December?

1.	2.	3.	4.
5.	6.	7.	8.
9.	10.	11.	12.
13.	14.	15.	16.
17.	18.	19.	20.

Lesson #9

1. 12.005 $\bigcirc$ 12.105

2. Find the difference of $\dfrac{3}{5}$ and $\dfrac{1}{2}$.

3. $(16 \div 4) \times 2 = ?$

4. Which digit is in the hundredths place in *7.9031*?

5. Find $\dfrac{2}{3}$ of 27.

6. Andy is 6 feet 6 inches tall. What is Andy's height in inches?

7. $13\dfrac{2}{5} + 15\dfrac{1}{4} = ?$

8. Find the perimeter of an equilateral triangle if a side is 25 mm.

9. $2 \div 3\dfrac{1}{2} = ?$

10. If $7x = 56$, what is the value of x?

11. $28.3 - 12.98 = ?$

12. How many pounds are in $6\dfrac{1}{2}$ tons?

13. $0.31 \times 0.06 = ?$

14. Nancy made 4 dozen cookies. If she ate $\dfrac{1}{6}$ of them, how many cookies did Nancy eat? How many are left?

15. If it is 2:45 now, what time will it be in 3 hours and 15 minutes?

16. $4.5 \div 6 = ?$

17. Write the formula for finding the area of a triangle.

18. How many millimeters are 5 meters?

19. $92 + 18 + 63 = ?$

20. Write the reciprocal of $\dfrac{3}{8}$.

1.	2.	3.	4.
5.	6.	7.	8.
9.	10.	11.	12.
13.	14.	15.	16.
17.	18.	19.	20.

Lesson #10

1. Which factors of 15 are also factors of 25?

2. Write *6.903* in words.

3. Name the shape to the right.

4. $(36 + 12) - (11 - 6) = ?$

5. Round *92.134* to the nearest tenth.

6. Numbers that have only two factors are _____ numbers.

7. Find the LCM of 16 and 18.

8. Find the area of the triangle.

9. $24\frac{1}{6} - 11\frac{5}{6} = ?$

10. Find $\frac{5}{6}$ of 30.

11. Make a factor tree for *45*.

12. $600 - 317 = ?$

13. $\frac{5}{6} \times \frac{2}{5} = ?$

14. What is the sum of the first five even whole numbers?

15. Explain how to find the average (mean) of a set of numbers.

16. Which digit is in the thousandths place in *5.807*?

17. What is the probability of rolling a seven on one throw of a die?

18. Arrange these decimals in order from greatest to least.

     4.7   4.007   4.87   4.07

19. When first planted, a sapling was 40 inches tall. Two years later it was 5 feet 7 inches tall. By how much did the sapling grow?

20. $8,399 + 6,475 = ?$

1.	2.	3.	4.
5.	6.	7.	8.
9.	10.	11.	12.
13.	14.	15.	16.
17.	18.	19.	20.

Lesson #11

1. $7.02 \times 0.5 = ?$

2.   How many minutes are in 6 hours?

3. $0.42 \div 0.06 = ?$

4.   Draw a hexagon. Show a line of symmetry.

5.   Write the number *5.214* using words.

6.   What do you call a triangle with no congruent sides?

7. $20 - 7\dfrac{5}{7} = ?$

8.   Find the GCF of 12 and 18.

9. $18\dfrac{3}{10} + 10\dfrac{2}{5} = ?$

10. $61,723 + 35,582 = ?$

11. $\dfrac{4}{5} \times \dfrac{15}{20} = ?$

12. Write $\dfrac{31}{5}$ as a mixed number.

13. How many years are in 6 decades?

14. Find the area of the parallelogram.

15. Round *53.091* to the nearest tenth.

16. Find the average of 16, 18, 20 and 22.

17. A triangle with two congruent sides is called _____.

18. Willie worked 7.5 hours on Saturday and 8.5 hours on Sunday. If he earns $7.25 per hour, how much did he earn on the weekend?

19. What is the probability of an event that is certain to happen?

20. Write the formula for finding the area of a triangle.

1.	2.	3.	4.
5.	6.	7.	8.
9.	10.	11.	12.
13.	14.	15.	16.
17.	18.	19.	20.

Lesson #12

1. Draw a ray.

2. Which is greater, 3 miles or 21,120 feet?

3. How many centimeters are in 12 meters?

4. $95,316 + 547,136 = ?$

5. Closed figures made up of line segments are _____.

6. Find the median and the mode of 131, 148, 214, 188 and 214.

7. Half of the diameter of a circle is called the _____.

8. $36\frac{3}{10} + 53\frac{1}{5} = ?$

9. Find the area of the triangle.

10. Find $\frac{4}{5}$ of 50.

18 m

22 m

11. What is the distance around the outside of a circle called?

12. Round *84,649,133* to the nearest hundred thousand.

13. A number with only two factors is a _____ number.

14. Round *84.621* to the nearest tenth.

15. $800 - 412 = ?$

16. $0.075 \times 0.03 = ?$

17. Write *9.581* using words.

18. $1\frac{1}{4} \div 2\frac{1}{2} = ?$

19. $10.1 - 6.393 = ?$

20. In December 34.5 inches of snow fell in Cleveland. During January 28.25 inches fell and in February 17.75 inches of snow fell. What was Cleveland's total snowfall during those three months?

1.	2.	3.	4.
5.	6.	7.	8.
9.	10.	11.	12.
13.	14.	15.	16.
17.	18.	19.	20.

Lesson #13

1. The answer in a subtraction problem is called the _____.

2. $\dfrac{3}{10} \times \dfrac{5}{9} = ?$

3. The distance around the outside of a circle is called _____.

4. Find the LCM of 14 and 20.

5. $(20 \times 3) + (65 - 15) = ?$

6. How many ounces are in 6 pounds?

7. A triangle with all congruent sides is called _____.

8. $65\dfrac{1}{6} + 28\dfrac{2}{3} = ?$

9. Write the formula for finding the volume of a rectangular prism.

10. Which digit is in the thousandths place in *3.0145*?

11. $\dfrac{8}{10} \div \dfrac{4}{10} = ?$

12. How long is this segment?

13. Make a factor tree for *24*.

14. If Frankie earned $8 per hour, how much will he earn if he works $3\dfrac{1}{2}$ hours? $4\dfrac{1}{4}$ hours?

15. Find the area of the parallelogram.

28 mm
34 mm

16. $16 - 9\dfrac{3}{8} = ?$

17. How many inches are in 3 feet?

18. $56.99 + 22.4 = ?$

19. Find the GCF of 12 and 16.

20. Figures with the same size and shape are _____.

1.	2.	3.	4.
5.	6.	7.	8.
9.	10.	11.	12.
13.	14.	15.	16.
17.	18.	19.	20.

Lesson #14

1. What kind of angle measures less than 90°? Draw one.

2. $\dfrac{5}{6} \times \dfrac{3}{5} = ?$

3. Estimate the product of 76 and 43.

4. How many tons are 16,000 pounds?

5. Write *nine and five thousandths* in standard decimal form.

6. Mrs. Smith is taking herself and 12 students to a movie. The cost of each ticket was $4.25. How much did she spend on the tickets?

7. $\dfrac{5}{9} \bigcirc \dfrac{7}{10}$

8. Round *4.583* to the nearest tenth.

9. $12\dfrac{2}{5} + 9\dfrac{1}{3} = ?$

10. Put $\dfrac{10}{15}$ in simplest form.

11. $75 + 83 + 91 = ?$

12. How many quarts are in 5 gallons?

13. Find the area of Maria's bedroom if it is 13 ft. long and 14 ft. wide.

14. $316 - 177 = ?$

15. List the factors of 24.

16. Draw intersecting lines.

17. Find the circumference of this circle.

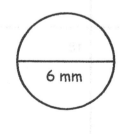

6 mm

18. Is the number *29* prime or composite?

19. Order these decimals from least to greatest.

    6.41   6.4   6.04   6.341

20. What is the top number in a fraction called?

1.	2.	3.	4.
5.	6.	7.	8.
9.	10.	11.	12.
13.	14.	15.	16.
17.	18.	19.	20.

Lesson #15

1. How many minutes are in 4 hours?

2. $9 - 3\dfrac{4}{7} = ?$

3. Are these figures similar or congruent?

4. Find the GCF of 12 and 18.

5. A baker used 480 boxes of cake mix to make a wedding cake. If each box called for 3 eggs, how many dozen eggs were used to make the wedding cake?

6. What is the probability of rolling a three on one throw of a die?

7. $0.9 - 0.441 = ?$

8. How many yards are in a mile?

9. $0.945 \div 0.5 = ?$

10. Find $\dfrac{2}{9}$ of 81.

11. $32 \times 65 = ?$

12. Write $\dfrac{8}{12}$ in simplest terms.

13. $4,516 + 2,918 = ?$

14. Draw perpendicular lines.

15. $(12 \times 6) \div 8 = ?$

16. What is the area of the parallelogram?

17. Round *24.817* to the nearest hundredth.

18. Convert 240 ounces into pounds.

19. What is the distance around the outside of a circle called?

20. The answer to a division problem is the _____.

1.	2.	3.	4.
5.	6.	7.	8.
9.	10.	11.	12.
13.	14.	15.	16.
17.	18.	19.	20.

Lesson #16

1.   A triangle with two congruent sides is a(n) _____ triangle.

2. $0.396 \bigcirc 0.42$

3.   Make a factor tree for *90*.

4. $0.007 \times 0.008 = ?$

5.   Write the formula for finding the circumference of a circle.

6.   Find the area of the triangle.

7.   Find $\dfrac{2}{5}$ of 40.

16 in.

22 in.

8. $25 \times 42 = ?$

9.   If the quotient is 6 and the dividend is 42, what is the divisor?

10. How many centimeters are in 5 meters?

11. $3.6 \div 0.6 = ?$

12. State whether the figure was moved by a rotation, a slide, or a reflection.

13. $31\dfrac{1}{6} - 14\dfrac{5}{6} = ?$

14. Write $\dfrac{59}{6}$ as a mixed number.

15. Round *1.918* to the nearest hundredth.

16. $2\dfrac{1}{4} \times 3\dfrac{1}{9} = ?$

17. $800 - 233 = ?$

18. Which factors of 12 are also factors of 24?

19. $\dfrac{8}{10} \div \dfrac{2}{5} = ?$

20. Mr. DiPalma has a board that is $12\dfrac{3}{5}$ ft. long. If he cuts the board into sections that are $1\dfrac{1}{5}$ ft. each, how many sections will he have?

1.	2.	3.	4.
5.	6.	7.	8.
9.	10.	11.	12.
13.	14.	15.	16.
17.	18.	19.	20.

Lesson #17

1. $0.8 - 0.2315 = ?$

2. What is the name of this shape?

3. Draw a cube. How many faces does it have?

4. If it is 5:20 now, what time was it 3 hours and 10 minutes ago?

5. How many centuries are 700 years?

6. $\dfrac{5}{8} \times \dfrac{4}{5} = ?$

7. Write $5\dfrac{2}{5}$ as an improper fraction.

8. $39 - 22\dfrac{5}{8} = ?$

9. $0.003 \times 0.04 = ?$

10. Find the average of 2.4, 6.3 and 5.7.

11. Round *176,438,913* to the nearest ten million.

12. $155,432 + 622,518 = ?$

13. Find the circumference of this circle.

14. Find the LCM of 10 and 15.

18 mm

15. What is the distance across the middle of a circle called?

16. If $9x = 72$, what is the value of x?

17. Write the formula for finding the area of a parallelogram.

18. $5,845 \div 6 = ?$

19. Find the probability of rolling a prime number on 1 roll of a die.

20. Myron has to bus tables at a country club party. He is scheduled to work for $1\dfrac{3}{4}$ hours. Myron will spend $\dfrac{2}{3}$ of that time washing dishes. How much time will he be washing dishes?

1.	2.	3.	4.
5.	6.	7.	8.
9.	10.	11.	12.
13.	14.	15.	16.
17.	18.	19.	20.

Lesson #18

1.   What do you call the distance around the outside of a circle?

2.   Find the GCF of 18 and 24.

3.   The area of a square is 49 ft². What is the length of each side?

4.   What time will it be in 9 hours and 25 minutes, if it is 5:10 now?

5. $\frac{7}{8} + \frac{1}{2} = ?$

6.   Draw a ray.

7. $18 - 9\frac{3}{7} = ?$

8.   How many millimeters are in 3 meters?

9. $8.5 \div 0.05 = ?$

10. Give the probability of rolling an even number on one roll of a die.

11. The radius of a circle is 32 centimeters. What is the diameter?

12. $0.9 - 0.3897 = ?$

13. Find the area of the parallelogram.

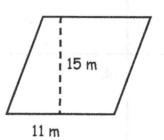
15 m
11 m

14. Find $\frac{1}{5}$ of 40.

15. Draw two similar pentagons.

16. Write *10.278* using words.

17. $\frac{4}{7} \times \frac{14}{16} = ?$

18. Find the median and the mode of 321, 286, 174, 418 and 286.

19. $(45 \div 9) + (12 + 6) = ?$

20. Marsha gets an employee discount at the store where she works. She saves $10 on each coat plus an extra $3.25 with today's sale. She also has another coupon for $5 off. If the coat she wants costs $95, how much does Marsha pay for her coat (with her discounts)?

1.	2.	3.	4.
5.	6.	7.	8.
9.	10.	11.	12.
13.	14.	15.	16.
17.	18.	19.	20.

Lesson #19

1. Make a factor tree for *30*.

2. $\dfrac{5}{11}$ ◯ $\dfrac{5}{8}$

3. Find the estimated sum of 7,321 and 3,974.

4. $28 \times 52 = ?$

5. Write the formula for finding the area of a triangle.

6. $1\dfrac{1}{2} \div \dfrac{2}{3} = ?$

7. Round *16.3412* to the nearest thousandth.

8. Write *7.218* in words.

9. Find the circumference of the circle.

12 cm

10. $60,000 - 23,894 = ?$

11. What number is next in this sequence?   135, 143, 151, _____

12. $0.093 \times 0.03 = ?$

13. Write $\dfrac{58}{9}$ as a mixed number.

14. $11\dfrac{3}{7} + 14\dfrac{1}{3} = ?$

15. How many cups are in 12 pints?

16. 0.5 ◯ 0.53

17. The area of a rectangle is 75 cm². If it is 15 cm long, how wide is it?

18. Draw a right angle. How many degrees are in a right angle?

19. Which is greater, 8 minutes or 360 seconds?

20. Thomas goes to summer camp every year. If his parents drive 135 miles to camp, drop him off, and then go back to pick him up in 2 weeks, how many miles will they have traveled?

1.	2.	3.	4.
5.	6.	7.	8.
9.	10.	11.	12.
13.	14.	15.	16.
17.	18.	19.	20.

Lesson #20

1. $56.40 \div 0.04 = ?$

2. Find the median of 336, 317, 329, 373 and 354.

3. Find $\frac{3}{5}$ of 15.

4. Find the area of the triangle.

5. $0.88 \times 0.04 = ?$

6. How many inches are in 4 yards?

7. $43 \times 52 = ?$

8. Round *71.65* to the nearest tenth.

9. Is this figure a polygon?

10. $\frac{5}{8} \times \frac{12}{20} = ?$

11. Find the LCM of 14 and 20.

12. $10 - 2\frac{5}{7} = ?$

13. If it is 2:25 now, what time was it 5 hours and 10 minutes ago?

14. $(72 \div 9) + 12 = ?$

15. Find the perimeter of the octagon.

16. $\frac{9}{10} \bigcirc \frac{7}{11}$

17. Write *5.216* in words.

18. Write the ratio *3:5* two other ways.

19. Find the perimeter of the irregular shape.

20. $575 \div 5 = ?$

1.	2.	3.	4.
5.	6.	7.	8.
9.	10.	11.	12.
13.	14.	15.	16.
17.	18.	19.	20.

Lesson #21

1. Draw perpendicular lines.

2. $0.05 \bigcirc 0.053$

3. Find the area of a parallelogram if its base is 16 mm and its height is 8 mm.

4. $0.6 - 0.1224 = ?$

5. Find the probability of rolling an odd number on one roll of a die.

6. $722,916 + 642,184 = ?$

7. What is the name of this shape?

8. Write $\dfrac{44}{5}$ as a mixed number.

9. $5.16 \times 0.04 = ?$

10. How many centuries are 900 years?

11. $21\dfrac{3}{5} + 17\dfrac{1}{4} = ?$

12. $\dfrac{7}{8} \times \dfrac{12}{21} = ?$

13. How many years are 10 decades?

14. Round *316,321,199* to the nearest hundred million.

15. Find the average of 63, 47 and 70.

16. What is the distance around the outside of a circle called?

17. Write $9\dfrac{2}{9}$ as an improper fraction.

18. $45.15 \div 0.3 = ?$

19. Find the mode and the range of 612, 523, 575, 612 and 566.

20. The cost of a rare coin was $2,000 in 1995, $3,250 in 2000, and $4,500 in 2005. If this pattern continues, what will the price likely be in 2010?

1.	2.	3.	4.
5.	6.	7.	8.
9.	10.	11.	12.
13.	14.	15.	16.
17.	18.	19.	20.

Lesson #22

1.   Figures having the same shape, but different sizes are _____.

2. $\dfrac{6}{8} \times \dfrac{4}{12} = ?$

3.   Find the GCF and LCM of 10 and 12.

4.   If the diameter of a circle is 50 inches, what is the radius?

5.   Write the ratio *2:3* two other ways.

6. $0.003 \times 0.006 = ?$

7.   Write the time 4 minutes before noon.

8.   Find $\dfrac{2}{3}$ of 27.

9.   Draw an acute angle.

10. Find the circumference of the circle.

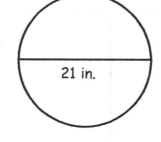

21 in.

11. $861 - 522 = ?$

12. How many hours are 420 minutes?

13. $1\dfrac{2}{3} \div \dfrac{2}{3} = ?$

14. The ratio of boys to girls in the classroom is 2 to 7. If there were 28 girls, how many boys are in the class?

15. How many pounds are $8\dfrac{1}{2}$ tons?

16. List the factors of *18*.

17. Find the volume of this rectangular prism.

6 m   12 m   3 m

18. Put these decimals in order from greatest to least.

        8.031   8.3   8.013   8.04

19. Which digit is in the millions place in *96,807,598?*

20. Is *28* a prime or a composite number?

44

1.	2.	3.	4.
5.	6.	7.	8.
9.	10.	11.	12.
13.	14.	15.	16.
17.	18.	19.	20.

Lesson #23

1. Draw a straight angle. How many degrees are in a straight angle?

2. $4\dfrac{1}{2} \times 2\dfrac{1}{3} = ?$

3. Round *31.279* to the nearest hundredth.

4. $\dfrac{5}{7} \times \dfrac{14}{25} = ?$

5. Name the solid figure that is shown?

6. $28.14 \div 0.2 = ?$

7. Write *5.7129* in words.

8. Write the ratio *4 to 9* in two other ways.

9. A football team of 30 players ate 15 pizzas. How many pizzas would be needed for 90 players?

10. $\dfrac{3}{4} \bigcirc \dfrac{5}{7}$

11. Solve this proportion for x. $\dfrac{5}{7} = \dfrac{x}{140}$

12. Draw parallel lines.

13. $\dfrac{8}{10} \div \dfrac{2}{10} = ?$

14. Find the LCM of 9 and 12.

15. How many quarts are in 8 gallons?

16. On the Celsius temperature scale, water boils at _____.

17. Write *seven and thirteen hundredths* in decimal form.

18. $52.88 + 9.319 = ?$

19. Closed figures made up of line segments are _____.

20. $0.006 \bigcirc 0.066$

1.	2.	3.	4.
5.	6.	7.	8.
9.	10.	11.	12.
13.	14.	15.	16.
17.	18.	19.	20.

Lesson #24

1. Numbers that have only 2 factors are _____ numbers.

2. $70,000 - 32,188 = ?$

3. On the Fahrenheit scale, water freezes at _____.

4. $5,765 \times 5 = ?$

5. Solve the proportion for x. $\dfrac{8}{12} = \dfrac{6}{x}$

6. An eight-sided shape is a(n) _____.

7. What will be the time 25 minutes before noon?

8. $31\dfrac{2}{7} - 12\dfrac{6}{7} = ?$

9. Draw a right angle. How many degrees are in a right angle?

10. $5.06 \bigcirc 5.006$

11. Find the GCF of 8 and 10.

12. $0.005 \times 0.09 = ?$

13. Find $\dfrac{2}{5}$ of 35.

14. Find the average of 65, 30 and 25.

15. $30\dfrac{4}{5} + 18\dfrac{1}{2} = ?$

16. What is the probability of rolling a 6 on one roll of a die?

17. Which digit is in the tenths place in *81.053*?

18. $\dfrac{3}{7} \times \dfrac{14}{18} = ?$

19. Write the formula for finding the circumference of a circle.

20. The seedling grows at the rate of 0.03 inch per day. At this rate, how much, in inches, will the seedling grow in 12 days?

1.	2.	3.	4.
5.	6.	7.	8.
9.	10.	11.	12.
13.	14.	15.	16.
17.	18.	19.	20.

Lesson #25

1. Draw intersecting lines.

2. How many feet are in 3 miles?

3. $19 - 6\dfrac{2}{5} = ?$

4. On the Celsius scale, at what temperature does water freeze?

5. Make a factor tree for *16*.

6. List the first 5 prime numbers.

7. Find the area of this parallelogram.

8. $\dfrac{8}{9} \times \dfrac{3}{12} = ?$

19 m

15 m

9. A triangle with all congruent sides is a(n) _____ triangle.

10. $0.7 - 0.319 = ?$

11. Write $\dfrac{19}{3}$ as a mixed number.

12. Write the reciprocal of $\dfrac{2}{9}$.

13. $\dfrac{9}{12} = \dfrac{x}{60}$

14. List the factors of 15.

15. Round *10.287* to the nearest hundredth.

16. Find the area of a square if a side measures 10 meters.

17. $87 + 23 + 41 = ?$

18. Write the ratio *5 to 8* two other ways.

19. Rewrite $\dfrac{14}{21}$ in simplest form.

20. The ratio of blue jays to robins was 5 to 7. If there were 84 robins, how many blue jays were there?

1.	2.	3.	4.
5.	6.	7.	8.
9.	10.	11.	12.
13.	14.	15.	16.
17.	18.	19.	20.

Lesson #26

1. Round *835,174,019* to the nearest hundred million.

2. $\dfrac{5}{8} - \dfrac{1}{3} = ?$

3. It is 1:30 now. What time will it be in 90 minutes?

4. The ratio of trucks to cars on a small stretch of highway was 5 to 12. If there were 85 trucks on the highway, how many cars were there?

5. If $4x = 60$, what is the value of x?

6. Find $\dfrac{3}{4}$ of 32.

7. Find the area of a rectangle with a length of 21 m and a width of 9 m.

8. How many decades are in 80 years?

9. Find the value of x. $\dfrac{5}{8} = \dfrac{60}{x}$

10. $18 - 10\dfrac{3}{4} = ?$

11. Which is greater, 5,000 pounds or 2 tons?

12. Write *13.86* in words.

13. $42.8 + 95.66 = ?$

14. Is *30* a prime or a composite number?

15. Draw an obtuse angle.

16. Find the circumference of the circle.

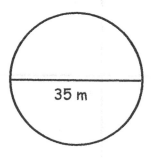

35 m

17. $2.8 \times 3.2 = ?$

18. Write the ratio $\dfrac{5}{7}$ two other ways.

19. Reduce $\dfrac{25}{30}$ to simplest terms.

20. Write *sixty-five and two hundred thirty-one thousandths* as a decimal.

1.	2.	3.	4.
5.	6.	7.	8.
9.	10.	11.	12.
13.	14.	15.	16.
17.	18.	19.	20.

Lesson #27

1. Write *2.43* in words.

2. On the Fahrenheit scale, water boils at _____.

3. $40\frac{1}{8} - 22\frac{7}{8} = ?$

4. $16.04 \div 0.04 = ?$

5. Which digit is in the thousandths place in *26.053*?

6. $300 - 112 = ?$

7. Solve the proportion for *x*. $\frac{3}{5} = \frac{x}{55}$

8. $9.2 - 4.995 = ?$

9. $15\frac{3}{5} + 31\frac{1}{4} = ?$

10. How many cups are in 10 pints?

11. $0.015 \times 0.05 = ?$

12. Figures with the same size and shape are _____.

13. Write the ratio *6:7* two other ways.

14. Order these decimals from least to greatest. 3.8   3.008   3.805

15. Find the GCF and the LCM of 15 and 20.

16. $134,865 + 518,922 = ?$

17. Find the area of the parallelogram.

18. $\frac{7}{9} \times \frac{3}{21} = ?$

20 cm

12 cm

19. $(5 \times 3) + 35 = ?$

20. Theresa had 40 cupcakes. She gave $\frac{1}{5}$ of the cupcakes to her aunt, $\frac{2}{5}$ of them to her grandma, and kept the rest. How many cupcakes did each person get?

1.	2.	3.	4.
5.	6.	7.	8.
9.	10.	11.	12.
13.	14.	15.	16.
17.	18.	19.	20.

Lesson #28

1. A triangle with no congruent sides is a(n) _____ triangle.

2. How many centuries are between 1740 and 1940?

3. Give the name of this polygon.

4. Draw parallel lines.

5. $4.9 - 0.7757 = ?$

6. Find $\dfrac{4}{7}$ of 42.

7. $11\dfrac{2}{5} + 13\dfrac{7}{10} = ?$

8. $15 - 5\dfrac{5}{8} = ?$

9. Make a factor tree for *80*.

10. $15 \times 15 = ?$

11. $3 \div 7\dfrac{1}{2} = ?$

12. How many centimeters are in 12 meters?

13. If $6x = 72$, what is the value of x?

14. How many millimeters are 6 meters?

15. Write the formula for finding the area of a triangle.

16. At what Fahrenheit temperature does water boil?

17. Solve the proportion for x. $\dfrac{5}{9} = \dfrac{x}{189}$

18. $\dfrac{5}{7} \bigcirc \dfrac{4}{9}$

19. The boy-girl ratio at the dance was 12 to 15. If there were 60 boys, how many girls were at the dance?

20. The area of a square is 64 square feet. How long is each side?

1.	2.	3.	4.
5.	6.	7.	8.
9.	10.	11.	12.
13.	14.	15.	16.
17.	18.	19.	20.

Lesson #29

1. If it is 4:15 now, what time was it 6 hours and 5 minutes ago?

2. Write the name of this shape.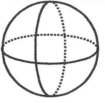

3. $13\frac{1}{5} - 4\frac{4}{5} = ?$

4. Write $8\frac{2}{3}$ as an improper fraction.

5. Closed figures made up of line segments are _____.

6. Explain how to find the median of a set of numbers.

7. Solve the proportion for x. $\frac{5}{9} = \frac{x}{135}$

8. Find the range of 112, 96, 83, 31 and 19.

9. Draw a ray.

10. $48 \div 0.08 = ?$

11. $90,000 - 55,416 = ?$

12. Write the ratio *5:13* two other ways.

13. Find the average of 1.3, 2.0 and 0.81.

14. $2\frac{1}{4} \div \frac{3}{4} = ?$

15. Find the area of the triangle.

16. $0.005 \times 0.07 = ?$

12 m

16 m

17. The ratio of kangaroo to elephants in the zoo is 5 to 3. If there were 25 kangaroo, how many elephants were in the zoo?

18. $3,819 \times 5 = ?$

19. A regular hexagon has a perimeter of 66 in. How long is each side?

20. $998 + 266 = ?$

1.	2.	3.	4.
5.	6.	7.	8.
9.	10.	11.	12.
13.	14.	15.	16.
17.	18.	19.	20.

Lesson #30

1. $0.007 \times 0.08 = ?$

2. Write the name of this solid shape.

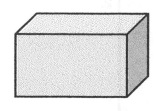

3. List the factors of *21*.

4. Round *31.781* to the nearest tenth.

5. What is the answer to a subtraction problem called?

6. $7,568 + 2,334 = ?$

7. Find $\dfrac{5}{6}$ of 60.

8. $\dfrac{5}{9} \times \dfrac{3}{5} = ?$

9. $0.3 - 0.1121 = ?$

10. How many inches are in 3 feet?

11. Which digit is in the thousandths place in *6.051*?

12. Write the ratio *7:11* in two other ways.

13. Round *84,320,557* to the nearest hundred thousand.

14. Solve the proportion for x. $\dfrac{2}{7} = \dfrac{x}{105}$

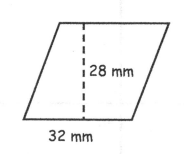

15. Find the area of the parallelogram.

16. Find the GCF of 10 and 12.

17. If $6n = 90$, what is the value of n?

18. What do you call the distance around the outside of a circle?

19. A triangle with 2 congruent sides is a(n) _____ triangle.

20. The ratio of tables to chairs in the banquet hall was 3 to 8. If there were 200 chairs, how many tables were in the hall?

1.	2.	3.	4.
5.	6.	7.	8.
9.	10.	11.	12.
13.	14.	15.	16.
17.	18.	19.	20.

Intermediate A

Mathematics

Help Pages &
"Who Knows?"

Help Pages

Vocabulary

Arithmetic operations

Difference — the result or answer to a subtraction problem. Example: The difference of 5 and 1 is 4.

Product — the result or answer to a multiplication problem. Example: The product of 5 and 3 is 15.

Quotient — the result or answer to a division problem. Example: The quotient of 8 and 2 is 4.

Sum — the result or answer to an addition problem. Example: The sum of 5 and 2 is 7.

Factors and Multiples

Factors — are multiplied together to get a product. Example: 2 and 3 are factors of 6.

Multiples — can be evenly divided by a number. Example: 5, 10, 15 and 20 are multiples of 5.

Composite Number — a number with more than 2 factors.
Example: 10 has factors of 1, 2, 5 and 10. Ten is a composite number.

Prime Number — a number with exactly 2 factors (the number itself and 1).
Example: 7 has factors of 1 and 7. Seven is a prime number.

Greatest Common Factor (GCF) — the highest factor that 2 numbers have in common.
Example: The factors of 6 are 1, 2, **3**, and 6. The factors of 9 are 1, **3** and 9. The GCF of 6 and 9 is **3**.

Least Common Multiple (LCM) — the smallest multiple that 2 numbers have in common.
Example: Multiples of 3 are 3, 6, 9, **12**, 15... Multiples of 4 are 4, 8, **12**, 16... The LCM of 3 and 4 is **12**.

Prime Factorization — a number, written as a product of its prime factors.
Example: 140 can be written as $2 \times 2 \times 5 \times 7$. (All of these are prime factors of 140.)

Fractions and Decimals

Improper Fraction — a fraction in which the numerator is larger than the denominator. Example: $\frac{9}{4}$

Mixed Number — the sum of a whole number and a fraction. Example: $5\frac{1}{4}$

Reciprocal — a fraction where the numerator and denominator are interchanged. The product of a fraction and its reciprocal is always 1.
Example: The reciprocal of $\frac{3}{5}$ is $\frac{5}{3}$. $\frac{3}{5} \times \frac{5}{3} = \frac{15}{15} = 1$

Repeating Decimal — a decimal in which a number or a series of numbers continues on and on.
Example: 2.33333333, 4.151515151515, 7.125555555, etc.

Geometry

Acute Angle — an angle measuring less than 90°.

Congruent — figures with the same shape and the same size.

Obtuse Angle — an angle measuring more than 90°.

Right Angle — an angle measuring exactly 90°.

Similar — figures having the same shape, but different size.

Straight Angle — an angle measuring exactly 180°.

Help Pages

Vocabulary (continued)

Geometry — Circles

Circumference — the distance around the outside of a circle.

Diameter — the widest distance across a circle. The diameter always passes through the center.

Radius — the distance from any point on the circle to the center. The radius is half of the diameter.

Geometry — Polygons

Number of Sides		Name	Number of Sides		Name
3	△	Triangle	7	⬡	Heptagon
4	▢	Quadrilateral	8	⬡	Octagon
5	⬠	Pentagon	9	⬡	Nonagon
6	⬡	Hexagon	10	⬡	Decagon

Geometry — Triangles

Equilateral — a triangle in which all 3 sides have the same length.

Isosceles — a triangle in which 2 sides have the same length.

Scalene — a triangle in which no sides are the same length.

Measurement — Relationships

Volume	Distance
3 teaspoons in a tablespoon	36 inches in a yard
2 cups in a pint	1760 yards in a mile
2 pints in a quart	5280 feet in a mile
4 quarts in a gallon	100 centimeters in a meter
Weight	1000 millimeters in a meter
16 ounces in a pound	**Temperature**
2000 pounds in a ton	0° Celsius – Freezing Point
Time	100°Celsius – Boiling Point
10 years in a decade	32°Fahrenheit – Freezing Point
100 years in a century	212°Fahrenheit – Boiling Point

Ratio and Proportion

Proportion — a statement that two ratios (or fractions) are equal. Example: $\frac{1}{2} = \frac{3}{6}$

Ratio — a comparison of two numbers by division; a ratio looks like a fraction. Example: $\frac{2}{5}$

Help Pages

Vocabulary (continued)

Statistics

Mean — the average of a group of numbers. The mean is found by finding the sum of a group of numbers and then dividing the sum by the number of members in the group.

Example: The average of 12, 18, 26, 17 and 22 is **19**. $\dfrac{12+18+26+17+22}{5} = \dfrac{95}{5} = 19$

Median — the middle value in a group of numbers. The median is found by listing the numbers in order from least to greatest, and finding the one that is in the middle of the list. If there is an even number of members in the group, the median is the average of the two middle numbers.

Example: The median of 14, 17, 24, 11 and 26 is **17**.   11, 14, ⓐ7, 24, 26

The median of 77, 93, 85, 95, 70 and 81 is **83**.   70, 77, 81, 85, 93, 95 $\dfrac{81+85}{2} = 83$

Mode — the number that occurs most often in a group of numbers. The mode is found by counting how many times each number occurs in the list. The number that occurs more than any other is the mode. Some groups of numbers have more than one mode.

Example: The mode of 77, 93, 85, 93, 77, 81, 93 and 71 is **93**.   (93 occurs more than the others.)

Place Value

Whole Numbers

$$8,\ 9\ 6\ 3,\ 2\ 7\ 1,\ 4\ 0\ 5$$

| Billions | Hundred Millions | Ten Millions | Millions | Hundred Thousands | Ten Thousands | Thousands | Hundreds | Tens | Ones |

The number above is read: eight billion, nine hundred sixty-three million, two hundred seventy-one thousand, four hundred five.

Decimal Numbers

$$1\ 7\ 8\ .\ 6\ 4\ 0\ 5\ 9\ 2$$

| Hundreds | Tens | Ones | Decimal Point | Tenths | Hundredths | Thousandths | Ten-thousandths | Hundred-thousandths | Millionths |

The number above is read: one hundred seventy-eight and six hundred forty thousand, five hundred ninety-two millionths.

Help Pages

Solved Examples

Factors and Multiples

The **Prime Factorization** of a number is when a number is written as a product of its prime factors. A factor tree is helpful in finding the prime factors of a number.

Example: Use a factor tree to find the prime factors of 45.

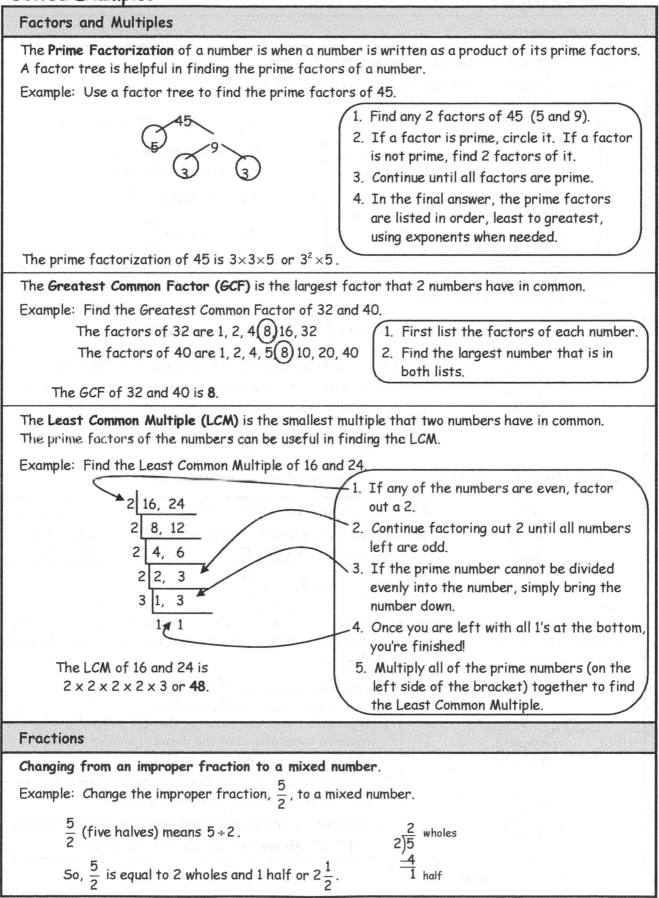

1. Find any 2 factors of 45 (5 and 9).
2. If a factor is prime, circle it. If a factor is not prime, find 2 factors of it.
3. Continue until all factors are prime.
4. In the final answer, the prime factors are listed in order, least to greatest, using exponents when needed.

The prime factorization of 45 is $3\times3\times5$ or $3^2\times5$.

The **Greatest Common Factor (GCF)** is the largest factor that 2 numbers have in common.

Example: Find the Greatest Common Factor of 32 and 40.

The factors of 32 are 1, 2, 4, 8, 16, 32
The factors of 40 are 1, 2, 4, 5, 8, 10, 20, 40

1. First list the factors of each number.
2. Find the largest number that is in both lists.

The GCF of 32 and 40 is **8**.

The **Least Common Multiple (LCM)** is the smallest multiple that two numbers have in common. The prime factors of the numbers can be useful in finding the LCM.

Example: Find the Least Common Multiple of 16 and 24.

```
2 | 16, 24
2 |  8, 12
2 |  4,  6
2 |  2,  3
3 |  1,  3
      1   1
```

1. If any of the numbers are even, factor out a 2.
2. Continue factoring out 2 until all numbers left are odd.
3. If the prime number cannot be divided evenly into the number, simply bring the number down.
4. Once you are left with all 1's at the bottom, you're finished!
5. Multiply all of the prime numbers (on the left side of the bracket) together to find the Least Common Multiple.

The LCM of 16 and 24 is
2 x 2 x 2 x 2 x 3 or **48**.

Fractions

Changing from an improper fraction to a mixed number.

Example: Change the improper fraction, $\frac{5}{2}$, to a mixed number.

$\frac{5}{2}$ (five halves) means $5\div2$.

So, $\frac{5}{2}$ is equal to 2 wholes and 1 half or $2\frac{1}{2}$.

$$\begin{array}{r} 2 \text{ wholes} \\ 2\overline{)5} \\ \underline{-4} \\ 1 \text{ half} \end{array}$$

Help Pages

Solved Examples

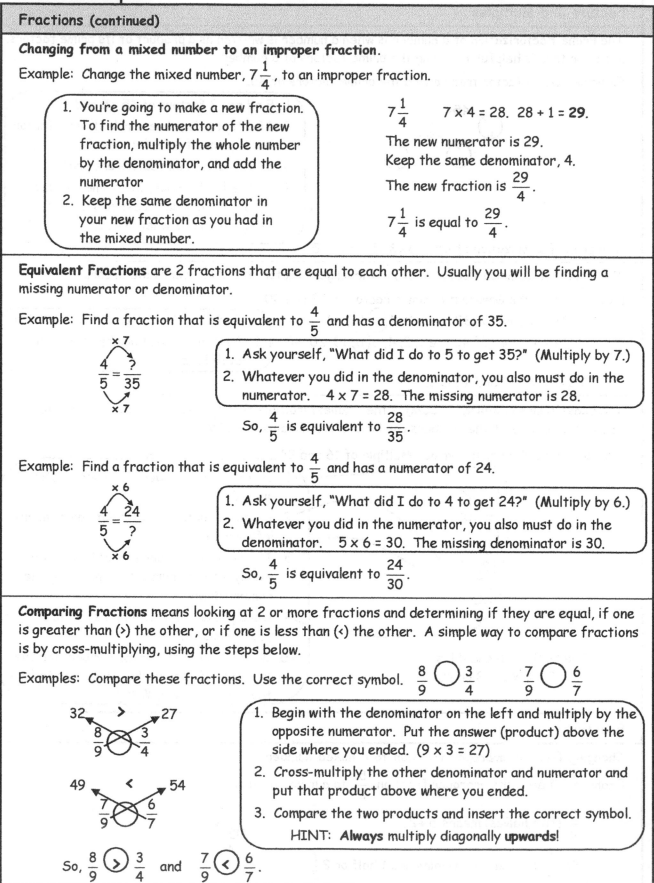

Fractions (continued)

Changing from a mixed number to an improper fraction.

Example: Change the mixed number, $7\frac{1}{4}$, to an improper fraction.

1. You're going to make a new fraction. To find the numerator of the new fraction, multiply the whole number by the denominator, and add the numerator
2. Keep the same denominator in your new fraction as you had in the mixed number.

$7\frac{1}{4}$ 7 x 4 = 28. 28 + 1 = **29**.

The new numerator is 29.
Keep the same denominator, 4.
The new fraction is $\frac{29}{4}$.

$7\frac{1}{4}$ is equal to $\frac{29}{4}$.

Equivalent Fractions are 2 fractions that are equal to each other. Usually you will be finding a missing numerator or denominator.

Example: Find a fraction that is equivalent to $\frac{4}{5}$ and has a denominator of 35.

$$\overset{\times\,7}{\underset{\times\,7}{\frac{4}{5}=\frac{?}{35}}}$$

1. Ask yourself, "What did I do to 5 to get 35?" (Multiply by 7.)
2. Whatever you did in the denominator, you also must do in the numerator.   4 x 7 = 28. The missing numerator is 28.

So, $\frac{4}{5}$ is equivalent to $\frac{28}{35}$.

Example: Find a fraction that is equivalent to $\frac{4}{5}$ and has a numerator of 24.

$$\overset{\times\,6}{\underset{\times\,6}{\frac{4}{5}=\frac{24}{?}}}$$

1. Ask yourself, "What did I do to 4 to get 24?" (Multiply by 6.)
2. Whatever you did in the numerator, you also must do in the denominator.   5 x 6 = 30. The missing denominator is 30.

So, $\frac{4}{5}$ is equivalent to $\frac{24}{30}$.

Comparing Fractions means looking at 2 or more fractions and determining if they are equal, if one is greater than (>) the other, or if one is less than (<) the other. A simple way to compare fractions is by cross-multiplying, using the steps below.

Examples: Compare these fractions. Use the correct symbol. $\frac{8}{9}\bigcirc\frac{3}{4}$ $\frac{7}{9}\bigcirc\frac{6}{7}$

$$32\;\overset{>}{\underset{}{\cancel{\frac{8}{9}\;\;\frac{3}{4}}}}\;27$$

1. Begin with the denominator on the left and multiply by the opposite numerator. Put the answer (product) above the side where you ended. (9 x 3 = 27)
2. Cross-multiply the other denominator and numerator and put that product above where you ended.
3. Compare the two products and insert the correct symbol.

   HINT: **Always** multiply diagonally **upwards**!

$$49\;\overset{<}{\underset{}{\cancel{\frac{7}{9}\;\;\frac{6}{7}}}}\;54$$

So, $\frac{8}{9}\;\bigcirc\!\!>\;\frac{3}{4}$ and $\frac{7}{9}\;\bigcirc\!\!<\;\frac{6}{7}$.

Help Pages

Solved Examples

Fractions (continued)	

To **add (or subtract) fractions with the same denominator**, simply add (or subtract) the numerators, keeping the same denominator.

Examples: $\dfrac{3}{5} + \dfrac{1}{5} = \dfrac{4}{5}$

$\dfrac{8}{9} - \dfrac{1}{9} = \dfrac{7}{9}$

To **add mixed numbers**, follow a process similar to the one you used with fractions. If the sum is an improper fraction, be sure to simplify it.

Example:
$$\begin{array}{r} 1\dfrac{2}{5} \\ +1\dfrac{4}{5} \\ \hline 2\dfrac{6}{5} \end{array}$$

$2\dfrac{6}{5}$ is improper. $\dfrac{6}{5}$ can be rewritten as $1\dfrac{1}{5}$.

So, $2\dfrac{6}{5}$ is $2 + 1\dfrac{1}{5} = 3\dfrac{1}{5}$.

When **adding fractions that have different denominators**, you need to change the fractions so they have a common denominator before they can be added.

Finding the **Least Common Denominator (LCD)**:

The LCD of the fractions is the same as the Least Common Multiple of the denominators. Sometimes, the LCD will be the product of the denominators.

Example: Find the sum of $\dfrac{3}{8}$ and $\dfrac{1}{12}$.

$$\begin{array}{r} \dfrac{3}{8} = \dfrac{9}{24} \\ +\dfrac{1}{12} = \dfrac{2}{24} \\ \hline \dfrac{11}{24} \end{array}$$

1. First, find the LCM of 8 and 12.
2. The LCM of 8 and 12 is 24. This is also the LCD of these 2 fractions.
3. Find an equivalent fraction for each that has a denominator of 24.
4. When they have a common denominator, the fractions can be added.

$$\begin{array}{l} 2\,\lfloor 8,12 \\ 2\,\lfloor 4,6 \\ 2\,\lfloor 2,3 \qquad 2\times2\times2\times3 = 24 \\ 3\,\lfloor 1,3 \\ \quad 1,1 \end{array}$$

The LCM is 24.
So, the LCD is 24.

Example: Add $\dfrac{1}{4}$ and $\dfrac{1}{5}$.

$$\begin{array}{r} \dfrac{1}{4} = \dfrac{5}{20} \\ +\dfrac{1}{5} = \dfrac{4}{20} \\ \hline \dfrac{9}{20} \end{array}$$

$4 \times 5 = 20$ The LCM is 20. Use 20 as the LCD.

When **adding mixed numbers with unlike denominators**, follow a process similar to the one you used with fractions (above). Be sure to put your answer in simplest form.

Example: Find the sum of $6\dfrac{3}{7}$ and $5\dfrac{2}{3}$.

$$\begin{array}{r} 6\dfrac{3}{7} = 6\dfrac{9}{21} \\ +5\dfrac{2}{3} = 5\dfrac{14}{21} \\ \hline 11\dfrac{23}{21} \\ \text{(improper)} \end{array}$$

$\dfrac{23}{21} = 1\dfrac{2}{21} + 11 = 12\dfrac{2}{21}$

1. Find the LCD.
2. Find the missing numerators.
3. Add the whole numbers, then add the fractions.
4. Make sure your answer is in simplest form.

Help Pages

Solved Examples

Fractions (continued)

When **subtracting numbers with unlike denominators**, follow a process similar to the one you used when adding fractions. Be sure to put your answer in simplest form.

Examples: Find the difference of $\frac{3}{4}$ and $\frac{2}{5}$.

$$\frac{3}{4} = \frac{15}{20}$$
$$-\frac{2}{5} = \frac{8}{20}$$
$$\frac{7}{20}$$

1. Find the LCD just as you did when adding fractions.
2. Find the missing numerators.
3. Subtract the numerators and keep the common denominator.
4. Make sure your answer is in simplest form.

Subtract $\frac{1}{16}$ from $\frac{3}{8}$.

$$\frac{3}{8} = \frac{6}{16}$$
$$-\frac{1}{16} = \frac{1}{16}$$
$$\frac{5}{16}$$

When **subtracting mixed numbers with unlike denominators**, follow a process similar to the one you used when adding mixed numbers. Be sure to put your answer in simplest form.

Example: Subtract $4\frac{2}{5}$ from $8\frac{9}{10}$.

1. Find the LCD.
2. Find the missing numerators.
3. Subtract and simplify your answer.

$$8\frac{9}{10} = 8\frac{9}{10}$$
$$-4\frac{2}{5} = 4\frac{4}{10}$$
$$4\frac{5}{10} = 4\frac{1}{2}$$

Sometimes when subtracting mixed numbers, you may need to regroup. If the numerator of the top fraction is smaller than the numerator of the bottom fraction, you must borrow from your whole number.

Example: Subtract $5\frac{5}{6}$ from $9\frac{1}{4}$.

1. Find the LCD.
2. Find the missing numerators.
3. Because you can't subtract 10 from 3, you need to borrow from the whole number.
4. Rename the whole number as a mixed number using the common denominator.
5. Add the 2 fractions to get an improper fraction.
6. Subtract the whole numbers and the fractions and simplify your answer.

$$9\frac{1}{4} = 9\frac{3}{12} = 8\frac{12}{12} + \frac{3}{12} = 8\frac{15}{12}$$
$$-5\frac{5}{6} = 5\frac{10}{12} = \qquad\qquad 5\frac{10}{12}$$
$$3\frac{5}{12}$$

More examples:

$$8\frac{1}{2} = 8\frac{2}{4} = 7\frac{4}{4} + \frac{2}{4} = 7\frac{6}{4}$$
$$-4\frac{3}{4} = 4\frac{3}{4} = \qquad\qquad 4\frac{3}{4}$$
$$3\frac{3}{4}$$

$$10\frac{1}{5} = 10\frac{4}{20} = 9\frac{20}{20} + \frac{4}{20} = 9\frac{24}{20}$$
$$-6\frac{3}{4} = 6\frac{15}{20} = \qquad\qquad 6\frac{15}{20}$$
$$3\frac{9}{20}$$

Help Pages

Solved Examples

Fractions (continued)

To **multiply fractions**, simply multiply the numerators together to get the numerator of the product. Then multiply the denominators together to get the denominator of the product. Make sure your answer is in simplest form.

Examples: Multiply $\frac{3}{5}$ by $\frac{2}{3}$.

$$\frac{3}{5} \times \frac{2}{3} = \frac{6}{15} = \frac{2}{3}$$

1. Multiply the numerators.
2. Multiply the denominators.
3. Simplify your answer.

Multiply $\frac{5}{8}$ by $\frac{4}{5}$.

$$\frac{5}{8} \times \frac{4}{5} = \frac{20}{40} = \frac{1}{2}$$

Sometimes you can use cancelling when multiplying fractions. Let's look at the examples again.

$$\frac{{}^{1}\cancel{3}}{5} \times \frac{2}{\cancel{3}_{1}} = \frac{2}{5}$$

The 3's have a common factor — 3. Divide both of them by 3. Since, $3 \div 3 = 1$, we cross out the 3's and write 1's in their place. Now, multiply the fractions. In the numerator, $1 \times 2 = 2$. In the denominator, $5 \times 1 = 5$.

The answer is $\frac{2}{5}$.

1. Are there any numbers in the numerator and the denominator that have common factors?
2. If so, cross out the numbers, divide both by that factor, and write the quotient.
3. Then, multiply the fractions as described above, using the quotients instead of the original numbers.

$$\frac{{}^{1}\cancel{5}}{{}_{2}\cancel{8}} \times \frac{\cancel{4}^{1}}{\cancel{5}_{1}} = \frac{1}{2}$$

As in the other example, the 5's can be cancelled. But here, the 4 and the 8 also have a common factor — 4. $8 \div 4 = 2$ and $4 \div 4 = 1$. After cancelling both of these, you can multiply the fractions.

REMEMBER: You can cancel up and down or diagonally, but NEVER sideways!

When **multiplying mixed numbers**, you must first change them into improper fractions.

Examples: Multiply $2\frac{1}{4}$ by $3\frac{1}{9}$.

$$2\frac{1}{4} \times 3\frac{1}{9} =$$

$$\frac{{}^{1}\cancel{9}}{{}_{1}\cancel{4}} \times \frac{\cancel{28}^{7}}{\cancel{9}_{1}} = \frac{7}{1} = 7$$

1. Change each mixed number to an improper fraction.
2. Cancel wherever you can.
3. Multiply the fractions.
4. Put your answer in simplest form.

Multiply $3\frac{1}{8}$ by 4.

$$3\frac{1}{8} \times 4 =$$

$$\frac{25}{{}_{2}\cancel{8}} \times \frac{\cancel{4}^{1}}{1} = \frac{25}{2} = 12\frac{1}{2}$$

To **divide fractions**, you must take the reciprocal of the 2nd fraction, and then multiply that reciprocal by the 1st fraction. Don't forget to simplify your answer!

Examples: Divide $\frac{1}{2}$ by $\frac{7}{12}$.

$$\frac{1}{2} \div \frac{7}{12} =$$

$$\frac{1}{\cancel{2}_{1}} \times \frac{\cancel{12}^{6}}{7} = \frac{6}{7}$$

1. Keep the 1st fraction as it is.
2. Write the reciprocal of the 2nd fraction.
3. Change the sign to multiplication.
4. Cancel if you can and multiply.
5. Simplify your answer.

Divide $\frac{7}{8}$ by $\frac{3}{4}$.

$$\frac{7}{8} \div \frac{3}{4} =$$

$$\frac{7}{{}_{2}\cancel{8}} \times \frac{\cancel{4}^{1}}{3} = \frac{7}{6} = 1\frac{1}{6}$$

Help Pages

Solved Examples

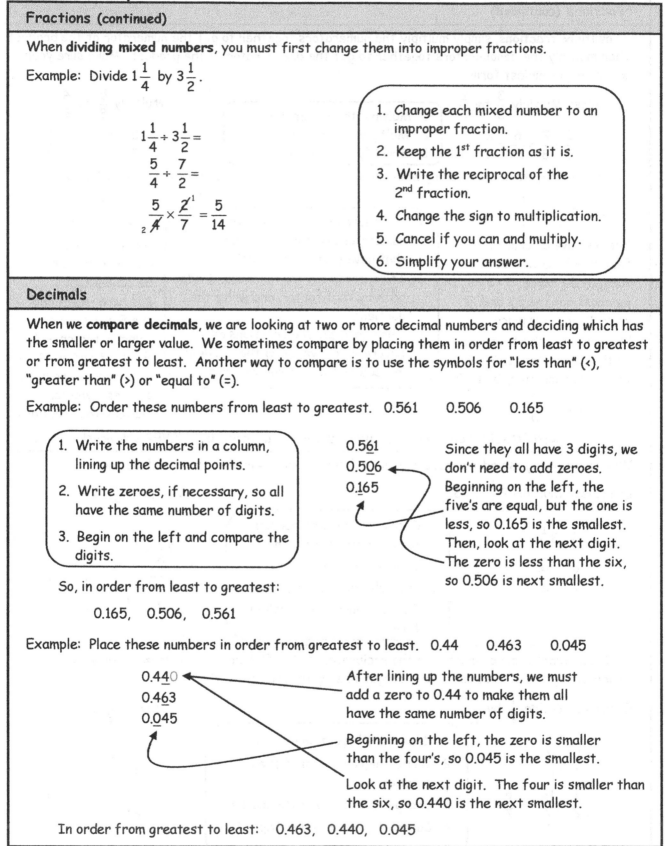

Fractions (continued)

When **dividing mixed numbers**, you must first change them into improper fractions.

Example: Divide $1\frac{1}{4}$ by $3\frac{1}{2}$.

$$1\frac{1}{4} \div 3\frac{1}{2} =$$
$$\frac{5}{4} \div \frac{7}{2} =$$
$$\frac{5}{\overset{2}{\cancel{4}}} \times \frac{\overset{1}{\cancel{2}}}{7} = \frac{5}{14}$$

1. Change each mixed number to an improper fraction.
2. Keep the 1st fraction as it is.
3. Write the reciprocal of the 2nd fraction.
4. Change the sign to multiplication.
5. Cancel if you can and multiply.
6. Simplify your answer.

Decimals

When we **compare decimals**, we are looking at two or more decimal numbers and deciding which has the smaller or larger value. We sometimes compare by placing them in order from least to greatest or from greatest to least. Another way to compare is to use the symbols for "less than" (<), "greater than" (>) or "equal to" (=).

Example: Order these numbers from least to greatest.   0.561   0.506   0.165

1. Write the numbers in a column, lining up the decimal points.
2. Write zeroes, if necessary, so all have the same number of digits.
3. Begin on the left and compare the digits.

0.5**6**1
0.5**0**6
0.**1**65

Since they all have 3 digits, we don't need to add zeroes. Beginning on the left, the five's are equal, but the one is less, so 0.165 is the smallest. Then, look at the next digit. The zero is less than the six, so 0.506 is next smallest.

So, in order from least to greatest:

   0.165,   0.506,   0.561

Example: Place these numbers in order from greatest to least.   0.44   0.463   0.045

0.4**4**0
0.4**6**3
0.**0**45

After lining up the numbers, we must add a zero to 0.44 to make them all have the same number of digits.

Beginning on the left, the zero is smaller than the four's, so 0.045 is the smallest.

Look at the next digit. The four is smaller than the six, so 0.440 is the next smallest.

In order from greatest to least: 0.463,   0.440,   0.045

Help Pages

Solved Examples

Decimals (continued)

When we **round decimals**, we are approximating them. This means we end the decimal at a certain place value and we decide if it's closer to the next higher number (round up) or to the next lower number (keep the same). It might be helpful to look at the decimal place-value chart on p. 286.

Example: Round 0.574 to the <u>tenths</u> place.

There is a 5 in the rounding (tenths) place.

0.5̲74

Since 7 is greater than 5, change the 5 to a 6.

0.5̲74

Drop the digits to the right of the tenths place.

0.6

1. Identify the number in the rounding place.

2. Look at the digit to its right.

3. If the digit is 5 or greater, increase the number in the rounding place by 1. If the digit is less than 5, keep the number in the rounding place the same.

4. Drop all digits to the right of the rounding place.

Example: Round 2.783 to the nearest <u>hundredth</u>.

2.78̲3   There is an 8 in the rounding place.

2.78̲3   Since 3 is less than 5, keep the rounding place the same

2.78   Drop the digits to the right of the hundredths place.

Adding and subtracting decimals is very similar to adding or subtracting whole numbers. The main difference is that you have to line-up the decimal points in the numbers before you begin.

Examples: Find the sum of 3.14 and 1.2.

$$\begin{array}{r} 3.14 \\ +\,1.20 \\ \hline 4.34 \end{array}$$

1. Line up the decimal points. Add zeroes as needed.

2. Add (or subtract) the decimals.

3. Add (or subtract) the whole numbers.

4. Bring the decimal point straight down.

Add 55.1, 6.472 and 18.33.

$$\begin{array}{r} 55.100 \\ 6.472 \\ +\,18.330 \\ \hline 79.902 \end{array}$$

Examples: Subtract 3.7 from 9.3.

$$\begin{array}{r} 9.3 \\ -\,3.7 \\ \hline 5.6 \end{array}$$

Find the difference of 4.1 and 2.88.

$$\begin{array}{r} 4.10 \\ -\,2.88 \\ \hline 1.22 \end{array}$$

Help Pages

Solved Examples

Decimals (continued)

When **multiplying a decimal by a whole number**, the process is similar to multiplying whole numbers.

Examples: Multiply 3.42 by 4.   Find the product of 2.3 and 2.

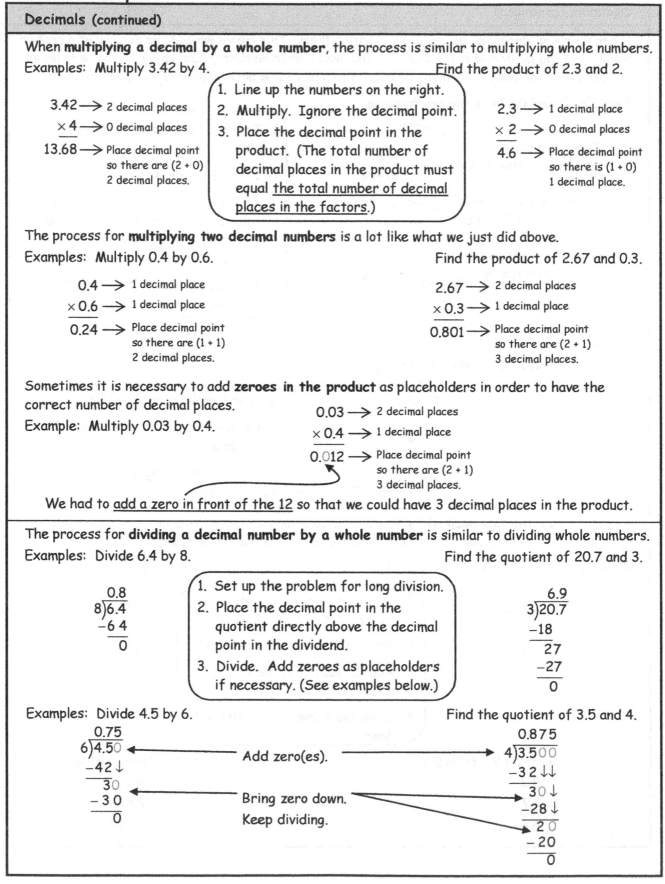

$3.42 \longrightarrow$ 2 decimal places
$\times 4 \longrightarrow$ 0 decimal places
$13.68 \longrightarrow$ Place decimal point so there are (2 + 0) 2 decimal places.

1. Line up the numbers on the right.
2. Multiply. Ignore the decimal point.
3. Place the decimal point in the product. (The total number of decimal places in the product must equal <u>the total number of decimal places in the factors</u>.)

$2.3 \longrightarrow$ 1 decimal place
$\times 2 \longrightarrow$ 0 decimal places
$4.6 \longrightarrow$ Place decimal point so there is (1 + 0) 1 decimal place.

The process for **multiplying two decimal numbers** is a lot like what we just did above.

Examples: Multiply 0.4 by 0.6.   Find the product of 2.67 and 0.3.

$0.4 \longrightarrow$ 1 decimal place
$\times 0.6 \longrightarrow$ 1 decimal place
$0.24 \longrightarrow$ Place decimal point so there are (1 + 1) 2 decimal places.

$2.67 \longrightarrow$ 2 decimal places
$\times 0.3 \longrightarrow$ 1 decimal place
$0.801 \longrightarrow$ Place decimal point so there are (2 + 1) 3 decimal places.

Sometimes it is necessary to add **zeroes in the product** as placeholders in order to have the correct number of decimal places.

Example: Multiply 0.03 by 0.4.

$0.03 \longrightarrow$ 2 decimal places
$\times 0.4 \longrightarrow$ 1 decimal place
$0.012 \longrightarrow$ Place decimal point so there are (2 + 1) 3 decimal places.

We had to <u>add a zero in front of the 12</u> so that we could have 3 decimal places in the product.

The process for **dividing a decimal number by a whole number** is similar to dividing whole numbers.

Examples: Divide 6.4 by 8.   Find the quotient of 20.7 and 3.

$$\begin{array}{r} 0.8 \\ 8\overline{)6.4} \\ -6\,4 \\ \hline 0 \end{array}$$

1. Set up the problem for long division.
2. Place the decimal point in the quotient directly above the decimal point in the dividend.
3. Divide. Add zeroes as placeholders if necessary. (See examples below.)

$$\begin{array}{r} 6.9 \\ 3\overline{)20.7} \\ -18 \\ \hline 27 \\ -27 \\ \hline 0 \end{array}$$

Examples: Divide 4.5 by 6.   Find the quotient of 3.5 and 4.

$$\begin{array}{r} 0.75 \\ 6\overline{)4.50} \\ -42\downarrow \\ \hline 30 \\ -30 \\ \hline 0 \end{array}$$

Add zero(es).

Bring zero down.
Keep dividing.

$$\begin{array}{r} 0.875 \\ 4\overline{)3.500} \\ -32\downarrow\downarrow \\ \hline 30\downarrow \\ -28\downarrow \\ \hline 20 \\ -20 \\ \hline 0 \end{array}$$

Help Pages

Solved Examples

Decimals (continued)

When dividing decimals the remainder is not always zero. Sometimes, the division continues on and on and the remainder begins to repeat itself. When this happens the quotient is called a **repeating decimal**.

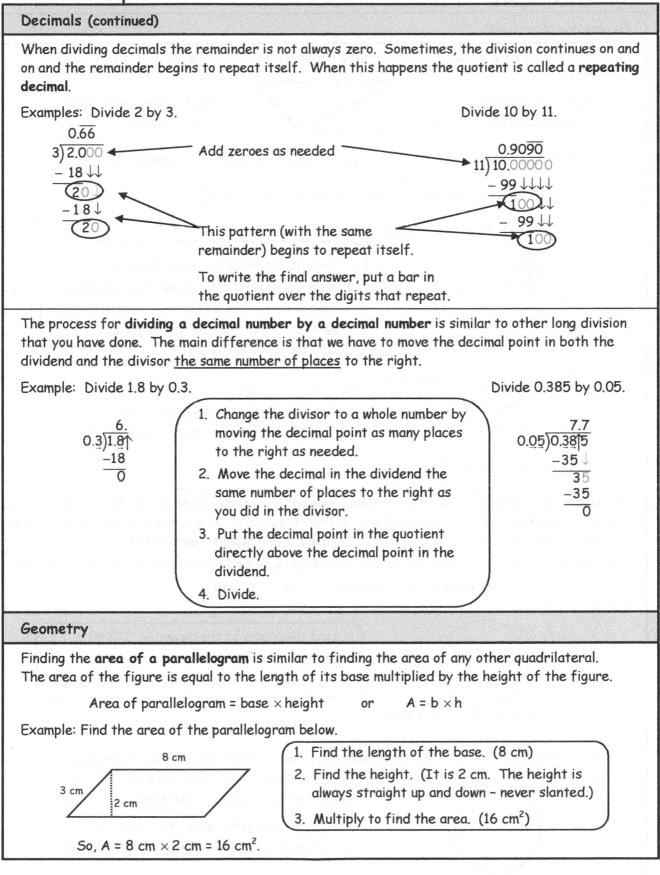

Examples: Divide 2 by 3.   Divide 10 by 11.

Add zeroes as needed

This pattern (with the same remainder) begins to repeat itself.

To write the final answer, put a bar in the quotient over the digits that repeat.

The process for **dividing a decimal number by a decimal number** is similar to other long division that you have done. The main difference is that we have to move the decimal point in both the dividend and the divisor <u>the same number of places</u> to the right.

Example: Divide 1.8 by 0.3.   Divide 0.385 by 0.05.

1. Change the divisor to a whole number by moving the decimal point as many places to the right as needed.

2. Move the decimal in the dividend the same number of places to the right as you did in the divisor.

3. Put the decimal point in the quotient directly above the decimal point in the dividend.

4. Divide.

Geometry

Finding the **area of a parallelogram** is similar to finding the area of any other quadrilateral. The area of the figure is equal to the length of its base multiplied by the height of the figure.

Area of parallelogram = base × height   or $A = b \times h$

Example: Find the area of the parallelogram below.

1. Find the length of the base. (8 cm)

2. Find the height. (It is 2 cm. The height is always straight up and down – never slanted.)

3. Multiply to find the area. (16 cm²)

So, $A = 8 \text{ cm} \times 2 \text{ cm} = 16 \text{ cm}^2$.

Help Pages

Solved Examples

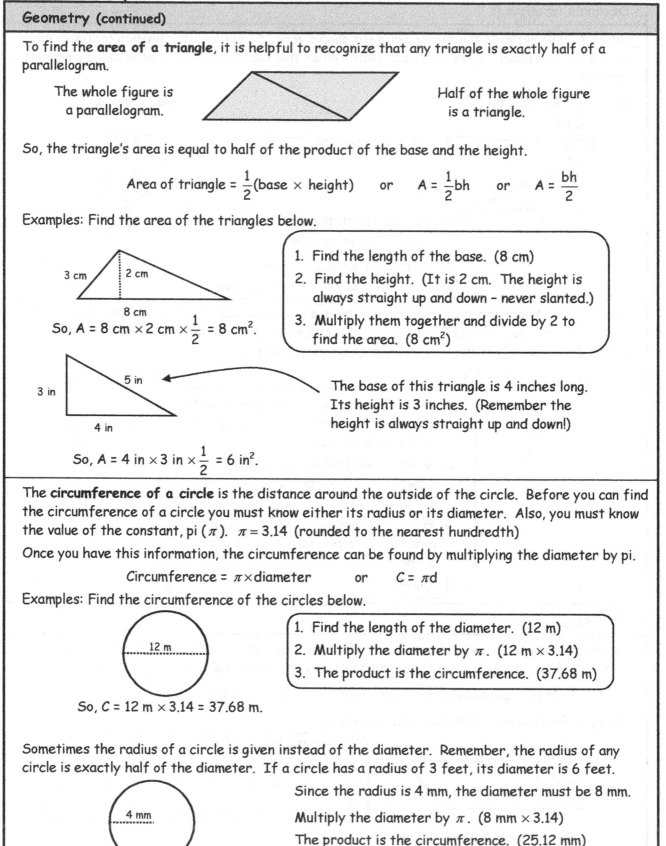

Geometry (continued)

To find the **area of a triangle**, it is helpful to recognize that any triangle is exactly half of a parallelogram.

The whole figure is a parallelogram.

Half of the whole figure is a triangle.

So, the triangle's area is equal to half of the product of the base and the height.

$$\text{Area of triangle} = \frac{1}{2}(\text{base} \times \text{height}) \quad \text{or} \quad A = \frac{1}{2}bh \quad \text{or} \quad A = \frac{bh}{2}$$

Examples: Find the area of the triangles below.

3 cm   2 cm   8 cm

So, $A = 8 \text{ cm} \times 2 \text{ cm} \times \frac{1}{2} = 8 \text{ cm}^2$.

1. Find the length of the base. (8 cm)
2. Find the height. (It is 2 cm. The height is always straight up and down – never slanted.)
3. Multiply them together and divide by 2 to find the area. (8 cm²)

5 in   3 in   4 in

The base of this triangle is 4 inches long. Its height is 3 inches. (Remember the height is always straight up and down!)

So, $A = 4 \text{ in} \times 3 \text{ in} \times \frac{1}{2} = 6 \text{ in}^2$.

The **circumference of a circle** is the distance around the outside of the circle. Before you can find the circumference of a circle you must know either its radius or its diameter. Also, you must know the value of the constant, pi (π). $\pi = 3.14$ (rounded to the nearest hundredth)

Once you have this information, the circumference can be found by multiplying the diameter by pi.

$$\text{Circumference} = \pi \times \text{diameter} \quad \text{or} \quad C = \pi d$$

Examples: Find the circumference of the circles below.

12 m

1. Find the length of the diameter. (12 m)
2. Multiply the diameter by π. (12 m × 3.14)
3. The product is the circumference. (37.68 m)

So, $C = 12 \text{ m} \times 3.14 = 37.68 \text{ m}$.

Sometimes the radius of a circle is given instead of the diameter. Remember, the radius of any circle is exactly half of the diameter. If a circle has a radius of 3 feet, its diameter is 6 feet.

4 mm

Since the radius is 4 mm, the diameter must be 8 mm.

Multiply the diameter by π. (8 mm × 3.14)

The product is the circumference. (25.12 mm)

So, $C = 8 \text{ mm} \times 3.14 = 25.12 \text{ mm}$.

Help Pages

Solved Examples

Ratio and Proportion

A **ratio** is used to compare two numbers. There are three ways to write a ratio comparing 5 and 7:

> 1. Word form ➡ 5 to 7
>
> 2. Fraction form ➡ $\dfrac{5}{7}$
>
> 3. Ratio form ➡ 5 : 7

You must make sure that all ratios are written in simplest form. (Just like fractions!!)

A **proportion** is a statement showing that two ratios are equal to each other. There are two ways to solve a proportion when a number is missing.

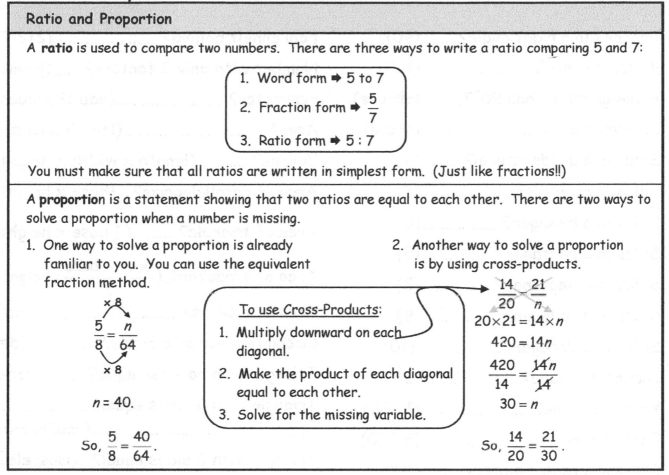

1. One way to solve a proportion is already familiar to you. You can use the equivalent fraction method.

$$\frac{5}{8} = \frac{n}{64}$$

×8 (top), ×8 (bottom)

$n = 40.$

So, $\dfrac{5}{8} = \dfrac{40}{64}$.

2. Another way to solve a proportion is by using cross-products.

To use Cross-Products:

1. Multiply downward on each diagonal.

2. Make the product of each diagonal equal to each other.

3. Solve for the missing variable.

$$\frac{14}{20} \quad \frac{21}{n}$$

$$20 \times 21 = 14 \times n$$

$$420 = 14n$$

$$\frac{420}{14} = \frac{14n}{14}$$

$$30 = n$$

So, $\dfrac{14}{20} = \dfrac{21}{30}$.

Who Knows???

Degrees in a right angle? ....(90)

A straight angle? ....(180)

Angle greater than 90°? ....(obtuse)

Less than 90°? ....(acute)

Sides in a quadrilateral? ....(4)

Sides in an octagon? ....(8)

Sides in a hexagon? ....(6)

Sides in a pentagon? ....(5)

Sides in a heptagon? ....(7)

Sides in a nonagon? ....(9)

Sides in a decagon? .... (10)

Inches in a yard? ....(36)

Yards in a mile? ....(1,760)

Feet in a mile? ....(5,280)

Centimeters in a meter? ....(100)

Teaspoons in a tablespoon? ...(3)

Ounces in a pound? ....(16)

Pounds in a ton?.......................(2,000)

Cups in a pint? ....(2)

Pints in a quart? ....(2)

Quarts in a gallon? ....(4)

Millimeters in a meter? ....(1,000)

Years in a century? ....(100)

Years in a decade? ....(10)

Celsius freezing? ....(0°C)

Celsius boiling? ....(100°C)

Fahrenheit freezing? ....(32°F)

Fahrenheit boiling? ....(212°F)

Number with only 2 factors? .....(prime)

Perimeter? ....(add the sides)

Area? ....(length x width)

Volume? ....(length x width x height)

Area of parallelogram? .. (base x height)

Area of triangle? ....($\frac{1}{2}$ base x height)

Area of trapezoid..($\frac{base + base}{2} \times$ height)

Area of a circle? ....(πr^2)

Circumference of a circle? ....(dπ)

Triangle with no sides equal? ..(scalene)

Triangle with 3 sides equal? ....
....(equilateral)

Triangle with 2 sides equal? .(isosceles)

Distance across the middle of a circle?
....(diameter)

Half of the diameter? ....(radius)

Figures with the same size and shape?
....(congruent)

Figures with same shape, different
sizes? ....(similar)

Number occurring most often? ...(mode)

Middle number? ....(median)

Answer in addition? ....(sum)

Answer in division? ....(quotient)

Answer in subtraction?........(difference)

Answer in multiplication? ....(product)

Intermediate A

Mathematics

Answers to Lessons

	Lesson #1		Lesson #2		Lesson #3
1	pentagon	1	1,166	1	thirteen and forty-five thousandths
2	18,075	2	68/7	2	17,352
3	76	3	6 tons	3	11:00
4	0	4	☐ ☐	4	30 years
5	33 r 20	5	$9\frac{7}{8}$	5	490,000,000
6	800,000,000	6	1, 2, 3, 6	6	95.04
7	180°	7	1.284	7	15.032 15.065 15.32 15.65
8	58,160	8	$6\frac{3}{20}$	8	76.81
9	2/3	9	add the sides	9	$7\frac{3}{4}$
10	0.4754	10	385	10	4
11	14	11	3	11	18
12	9.1	12	$x = 4$	12	0.1059
13	1, 2, 3, 6, 9, 18	13	2/5	13	4
14	0.00024	14	polygons	14	1,713
15	prime	15	$13\frac{2}{5}$	15	similar
16	9.004	16	57°	16	$37\frac{1}{3}$
17	$94\frac{13}{40}$	17	reflection	17	▱
18	congruent	18	72.84	18	1
19	0.044 0.158 0.175 0.6	19	diameter	19	>
20	$15	20	317	20	5,525 ounces

	Lesson #4		Lesson #5		Lesson #6
1		**1**	quadrilaterals	**1**	10 cups
2	$2^2 \times 3^2$	**2**	2.45 2.363 2.303 2.045	**2**	$4\dfrac{5}{6}$
3	66,781	**3**	5/12 1/3 0	**3**	$-10\,°F$
4	700 years	**4**	17,343	**4**	12.35
5	90°	**5**	$8\dfrac{2}{9}$	**5**	$8\dfrac{5}{7}$
6	five and two thousand seventy-one ten thousandths	**6**	12 quarts	**6**	108
7	3	**7**	$<$	**7**	prime
8	55.9	**8**	3/5	**8**	1/6
9	$8\dfrac{1}{2}$	**9**	0.0888	**9**	70.6
10	$7\dfrac{2}{9}$	**10**	21 tsp.	**10**	$126\ \text{in}^2$
11	27,183	**11**	26,744,000	**11**	0.000048
12	180 in.	**12**	congruent	**12**	$x = 9$
13	2.142	**13**	1	**13**	1/2
14	0.24	**14**	0.3439	**14**	5
15	$343\ \text{mm}^3$	**15**	$120\ \text{m}^2$	**15**	74,418
16	70	**16**	7.01	**16**	2×5^2
17	$1\dfrac{3}{20}$	**17**	1,153	**17**	$9\dfrac{1}{3}$
18	radius	**18**	10.003	**18**	$V = l \times w \times h$
19	$<$	**19**	$16\dfrac{1}{4}$ or 16 hours, 15 minutes	**19**	more than 90°
20	$600 Answers may vary.	**20**	2,418	**20**	$7.00

	Lesson #7		Lesson #8		Lesson #9
1	mode	1	1,948	1	<
2	48	2	8 mm	2	1/10
3	1	3	53.242	3	8
4	12	4	Average = 49 Mode = 52	4	0
5	<	5	$A = lw$	5	18
6	3	6	500	6	78 inches
7	1,200 cm	7	cylinder	7	$28\frac{13}{20}$
8	$A = bh$	8	22	8	75 mm
9	$P = 46$ in. $A = 90 \text{in}^2$	9	>	9	4/7
10	66/7	10	37.22	10	$x = 8$
11	parallelogram	11	1/2	11	15.32
12	3/5	12	12	12	13,000 lb.
13	20.12	13	15.1	13	0.0186
14	1.0686	14	26,400 ft.	14	8 cookies eaten 40 cookies left
15	212°F	15	$58\frac{1}{10}$	15	6:00
16	100°C	16	0.0035	16	0.75
17	1	17	48 cm^2	17	$A = \frac{1}{2}bh$
18	rotation	18		18	5,000 mm
19	slide	19	denominator	19	173
20	reflection	20	84 cars	20	8/3

	Lesson #10		Lesson #11		Lesson #12
1	1, 5	1	3.51	1	
2	six and nine hundred three thousandths	2	360 minutes	2	21,120 feet
3	parallelogram	3	7	3	1,200 cm
4	43	4	Any	4	642,452
5	92.1	5	five and two hundred fourteen thousandths	5	polygons
6	prime	6	a scalene triangle	6	Median = 188 Mode = 214
7	144	7	$12\frac{2}{7}$	7	radius
8	96 in^2	8	6	8	$89\frac{1}{2}$
9	$12\frac{1}{3}$	9	$28\frac{7}{10}$	9	198 m^2
10	25	10	97,305	10	40
11	$3^2 \times 5$	11	3/5	11	circumference
12	283	12	$6\frac{1}{5}$	12	84,600,000
13	1/3	13	60 years	13	prime
14	20	14	224 cm^2	14	84.6
15	Add the numbers to be averaged and divide by the total number of addends.	15	53.1	15	388
16	7	16	19	16	0.00225
17	0	17	an isosceles triangle	17	nine and five hundred eighty-one thousandths
18	4.87 4.7 4.07 4.007	18	$116	18	1/2
19	27 in. or 2 ft. 3 in.	19	1	19	3.707
20	14,874	20	$A = \frac{1}{2}bh$	20	80.5 inches

	Lesson #13		Lesson #14		Lesson #15
1	difference	**1**	acute angle	**1**	240 minutes
2	1/6	**2**	1/2	**2**	$5\frac{3}{7}$
3	circumference	**3**	3,200	**3**	congruent
4	140	**4**	8 tons	**4**	6
5	110	**5**	9.005	**5**	120 dozen
6	96 ounces	**6**	$55.25	**6**	1/6
7	an equilateral triangle	**7**	<	**7**	0.459
8	$93\frac{5}{6}$	**8**	4.6	**8**	1,760 yd.
9	$V = l \times w \times h$	**9**	$21\frac{11}{15}$	**9**	1.89
10	4	**10**	2/3	**10**	18
11	2	**11**	249	**11**	2,080
12	$1\frac{3}{4}$ in.	**12**	20 quarts	**12**	2/3
13	$2^3 \times 3$	**13**	182 ft^2	**13**	7,434
14	$28 $34	**14**	139	**14**	
15	952 mm^2	**15**	1, 2, 3, 4, 6, 8, 12, 24	**15**	9
16	$6\frac{5}{8}$	**16**		**16**	432 mm^2
17	36 in.	**17**	18.84 mm	**17**	24.82
18	79.39	**18**	prime	**18**	15 pounds
19	4	**19**	6.04 6.341 6.4 6.41	**19**	circumference
20	congruent	**20**	numerator	**20**	quotient

	Lesson #16		Lesson #17		Lesson #18
1	isosceles	1	0.5685	1	circumference
2	$<$	2	octagon	2	6
3	$2 \times 3^2 \times 5$	3	6 faces	3	7 ft.
4	0.000056	4	2:10	4	2:35
5	$d\pi$ or $2\pi r$	5	7 centuries	5	$1\dfrac{3}{8}$
6	176 in²	6	1/2	6	
7	16	7	27/5	7	$8\dfrac{4}{7}$
8	1,050	8	$16\dfrac{3}{8}$	8	3,000 mm
9	7	9	0.00012	9	170
10	500 cm	10	4.8	10	1/2
11	6	11	180,000,000	11	64 cm
12	reflection	12	777,950	12	0.5103
13	$16\dfrac{1}{3}$	13	56.52 mm	13	165 m²
14	$9\dfrac{5}{6}$	14	30	14	8
15	1.92	15	diameter	15	
16	7	16	$x = 8$	16	ten and two hundred seventy-eight thousandths
17	567	17	$A = bh$	17	1/2
18	1, 2, 3, 4, 6, 12	18	974 r 1	18	both are 286
19	2	19	1/2	19	23
20	$10\dfrac{1}{2}$	20	$1\dfrac{1}{6}$ hours or 1 hour, 10 minutes	20	$76.75

	Lesson #19		Lesson #20		Lesson #21
1	$2 \times 3 \times 5$	1	1,410	1	
2	<	2	336	2	<
3	11,000	3	9	3	128 mm^2
4	1,456	4	36 mm^2	4	0.4776
5	$A = \frac{1}{2}bh$	5	0.0352	5	1/2
6	$2\frac{1}{4}$	6	144 in.	6	1,365,100
7	16.341	7	2,236	7	parallelogram
8	seven and two hundred eighteen thousandths	8	71.7	8	$8\frac{4}{5}$
9	37.68 cm	9	no	9	0.2064
10	36,106	10	3/8	10	9 centuries
11	159	11	140	11	$38\frac{17}{20}$
12	0.00279	12	$7\frac{2}{7}$	12	1/2
13	$6\frac{4}{9}$	13	9:15	13	100 years
14	$25\frac{16}{21}$	14	20	14	300,000,000
15	24 cups	15	104 m	15	60
16	<	16	>	16	circumference
17	5 cm	17	five and two hundred sixteen thousandths	17	83/9
18	90°	18	$\frac{3}{5}$ 3 to 5	18	150.5
19	8 minutes	19	38 cm	19	mode = 612 range = 89
20	540 miles	20	115	20	$5,750

	Lesson #22		Lesson #23		Lesson #24
1	similar	1	◄─────► 180°	1	prime
2	1/4	2	$10\frac{1}{2}$	2	37,812
3	GCF = 2 LCM = 60	3	31.28	3	32°F
4	25 inches	4	2/5	4	28,825
5	2/3 2 to 3	5	cube	5	$x = 9$
6	0.000018	6	140.7	6	octagon
7	11:56 a.m.	7	five and seven thousand one hundred twenty-nine ten thousandths	7	11:35 a.m.
8	18	8	4/9 4:9	8	$18\frac{3}{7}$
9		9	45 pizzas	9	90°
10	65.94 in.	10	>	10	>
11	339	11	$x = 100$	11	2
12	7 hours	12		12	0.00045
13	$2\frac{1}{2}$	13	4	13	14
14	8 boys	14	36	14	40
15	17,000 pounds	15	32 quarts	15	$49\frac{3}{10}$
16	1, 2, 3, 6, 9, 18	16	100°C	16	1/6
17	216 m³	17	7.13	17	0
18	8.3 8.04 8.031 8.013	18	62.199	18	1/3
19	6	19	polygons	19	$d\pi$ or $2\pi r$
20	composite	20	<	20	0.36 in.

	Lesson #25		Lesson #26		Lesson #27
1		1	800,000,000	1	two and forty-three hundredths
2	15,840 ft.	2	7/24	2	212°F
3	$12\frac{3}{5}$	3	3:00	3	$17\frac{1}{4}$
4	0°C	4	204 cars	4	401
5	2^4	5	$x = 15$	5	3
6	2, 3, 5, 7, 11	6	24	6	188
7	285 m²	7	189 m²	7	$x = 33$
8	2/9	8	8 decades	8	4.205
9	equilateral	9	$x = 96$	9	$46\frac{17}{20}$
10	0.381	10	$7\frac{1}{4}$	10	20 cups
11	$6\frac{1}{3}$	11	5,000 lb.	11	0.00075
12	9/2	12	thirteen and eighty-six hundredths	12	congruent
13	$x = 45$	13	138.46	13	6/7 6 to 7
14	1, 3, 5, 15	14	composite	14	3.008 3.8 3.805
15	10.29	15		15	GCF = 5 LCM = 60
16	100 m²	16	109.9 m	16	653,787
17	151	17	8.96	17	240 cm²
18	5/8 5:8	18	5:7 5 to 7	18	1/9
19	2/3	19	5/6	19	50
20	60 blue jays	20	65.231	20	Theresa – 16 cupcakes aunt – 8 cupcakes grandma – 16 cupcakes

	Lesson #28		Lesson #29		Lesson #30
1	scalene	1	10:10	1	0.00056
2	2 centuries	2	sphere	2	rectangular prism
3	parallelogram	3	$8\frac{2}{5}$	3	1, 3, 7, 21
4	↔ ↔ ↕	4	26/3	4	31.8
5	4.1243	5	polygons	5	difference
6	24	6	Put the numbers in order, least to greatest and find the middle number.	6	9,902
7	$25\frac{1}{10}$	7	$x = 75$	7	50
8	$9\frac{3}{8}$	8	93	8	1/3
9	$2^4 \times 5$	9	⟶	9	0.1879
10	225	10	600	10	36 inches
11	2/5	11	34,584	11	1
12	1,200 cm	12	5/13 5 to 13	12	7/11 7 to 11
13	$x = 12$	13	1.37	13	84,300,000
14	6,000 mm	14	3	14	$x = 30$
15	$A = \frac{1}{2}bh$	15	96 m^2	15	896 mm^2
16	212°F	16	0.00035	16	2
17	$x = 105$	17	15 elephants	17	$n = 15$
18	>	18	19,095	18	circumference
19	75 girls	19	11 inches	19	isosceles
20	8 feet	20	1,264	20	75 tables